Alan McKirdy has written many popular books and book chapters on geology and related topics, and has helped to promote the study of environmental geology in Scotland. His other books with Birlinn include *Set in Stone: The Geology and Landscapes of Scotland* and *Land of Mountain and Flood*, which was nominated for the Saltire Research Book of the Year prize. *Northern Highlands – Landscapes in Stone* was long-listed for the Highland Book Prize. He is also the author of *James Hutton: The Founder of Modern Geology.* Before his retirement, Alan was Head of Knowledge and Information Management at Scottish Natural Heritage. Alan is now a freelance writer and has given many talks on Scottish geology and landscapes at book festivals and other events across the country.

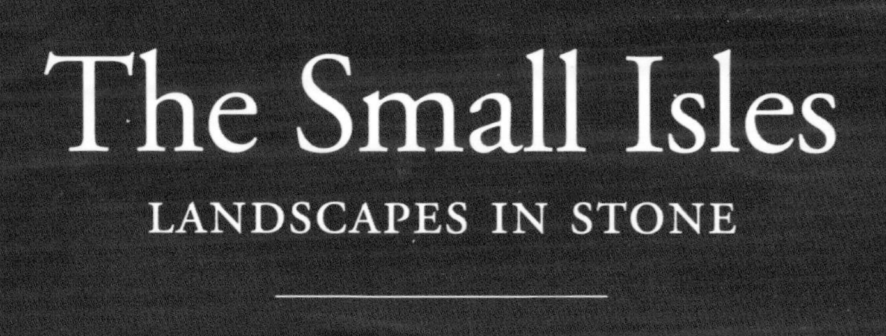

The Small Isles

LANDSCAPES IN STONE

Alan McKirdy

For Lorne Gill

First published in Great Britain in 2022 by
Birlinn Ltd
West Newington House
10 Newington Road
Edinburgh
EH9 1QS

www.birlinn.co.uk

ISBN: 978 178027 750 9

British Library Cataloguing-in-Publication Data
A catalogue record for this book is available on request from the British Library

Designed and typeset by Mark Blackadder

FRONTISPIECE:
Singing Sands Beach, Cleadale, Isle of Eigg.

Birlinn Ltd would like to thank

for their generous donation towards this publication.

Printed and bound by Gutenberg Press Ltd, Malta

Contents

Introduction

The Small Isles comprise the scenic Inner Hebridean islands of Rum, Eigg, Canna and Muck. Human occupation of these islands dates back some 9,000 years to Mesolithic times. Rum was home to the first human settlers, shortly after conditions warmed up at the end of the last Ice Age. Britain was joined to continental Europe by a land bridge until around 8,500 years ago, so nomadic tribes were free to wander wherever they chose. The earliest known settlement in the area, and perhaps in Scotland, was found near Kinloch on Rum, from where arrowheads, scrapers and blades have been recovered from the soil.

Opposite.
View of the mountainous southern part of the Isle of Rum.

The landscapes, rocks and fossils of these beautiful, remote islands tell of a drama even older than that, involving an ancient ecosystem populated by dinosaurs and, earlier still, a desert landscape. The geological history stretches back 3 billion years to some of the earliest events recorded on Planet Earth. From these far-off times, we can trace the geological events that gave rise to the landscapes we see today. That record of the rocks is incomplete, and fragmentary, with large gaps where no evidence remains. Nevertheless, the rocks and landscapes tell a compelling narrative of volcanoes belching molten magma and searing ash flows, interspersed with periods of tranquillity where tropical seas teemed with exotic life.

All four islands owe their origin to a group of three adjacent volcanoes that were active around 60 million years ago. Rum is the eroded remains of the magma chamber of one of these volcanoes. Eigg and Muck are part of the lava field that extends north from the Mull volcano, and Canna represents the southern extent of the lavas that flowed from the Skye volcano.

The next event that left its mark on these islands was the Ice Age. Starting around 2.4 million years ago, during this period of intense cold the ice sheets waxed and waned, according to the prevailing global temperatures. Its effect on the landscape was profound. The thick cover of erosive ice shaped the contours of the land into the hills and glens that we are familiar with today.

The Small Isles through time

Period of geological time	*Millions of years ago*	*Scotland's global position*	*Environments and events in the Small Isles Scotland*
Anthropocene	Last 10,000 years	57° N	People migrated to Rum around 9,000 years ago an have probably inhabited the Small Isles ever since.
Quaternary	Started 2.4 million years ago	Present position of 57° N	• **9,000 to 6,500 years ago** – sea levels rose and raised beaches were formed. • **11,500 onwards** – the ice retreated as the climate started to warm. • **12,500 to 11,500 years ago** – the climate became very cold as the ice returned: the Loch Lomond Readvance. • **29,000 to 14,700 years ago** – the landscape was entirely covered by an ice sheet during this, the last advance of the ice. • **Before 29,000 years ago** and for a period approaching the last 2 million years, there were prolonged periods when thick sheets of ice covered the area.
Neogene	23–2	55° N	Temperatures fell as the Ice Age approached.
Palaeogene	66–23	50° N	The bedrock of the Small Isles was largely formed during these turbulent volcanic times, with intermittent outpourings of lava.
Cretaceous	145–66	40° N	Sea levels rose to drown the area.
Jurassic	201–145	35° N	Thick sequences of Jurassic rocks are exposed on Eigg and Muck. An exciting find of a stegosaurus limb bone has recently been made on Eigg.

Period of geological time	*Millions of years ago*	*Scotland's global position*	*Environments and events in the Small Isles Scotland*
Triassic	252–201	30° N	Desert conditions prevailed and a small, isolated pocket of sandstone on Rum dates from this time.
Permian	299–252	20° N	Deserts covered the land, but there are no rocks of this age in the Small Isles.
Carboniferous	359–299	On the Equator	'Scotland' was located at the Equator at this time, but there are no rocks of this age in the Small Isles.
Devonian	419–359	10° S	Arid conditions with extensive river systems were widespread, but there are no rocks of this age in the Small Isles.
Silurian	444–419	15° S	Large upheavals created the Highlands of Scotland, but there are no rocks of this age in the Small Isles.
Ordovician	485–444	20° S	No rocks of this age are found in the Small Isles.
Cambrian	541–485	30° S	No rocks of this age are found in the Small Isles.
Proterozoic	2,500–541	Close to the South Pole	The Torridonian sandstones are dated around 900 million years old. They outcrop extensively on Rum.
Archaean	Prior to 2,500	Possibly close to the South Pole	Small patches of Lewisian gneiss are exposed on Rum. The age of the Earth is around 4,543 million years

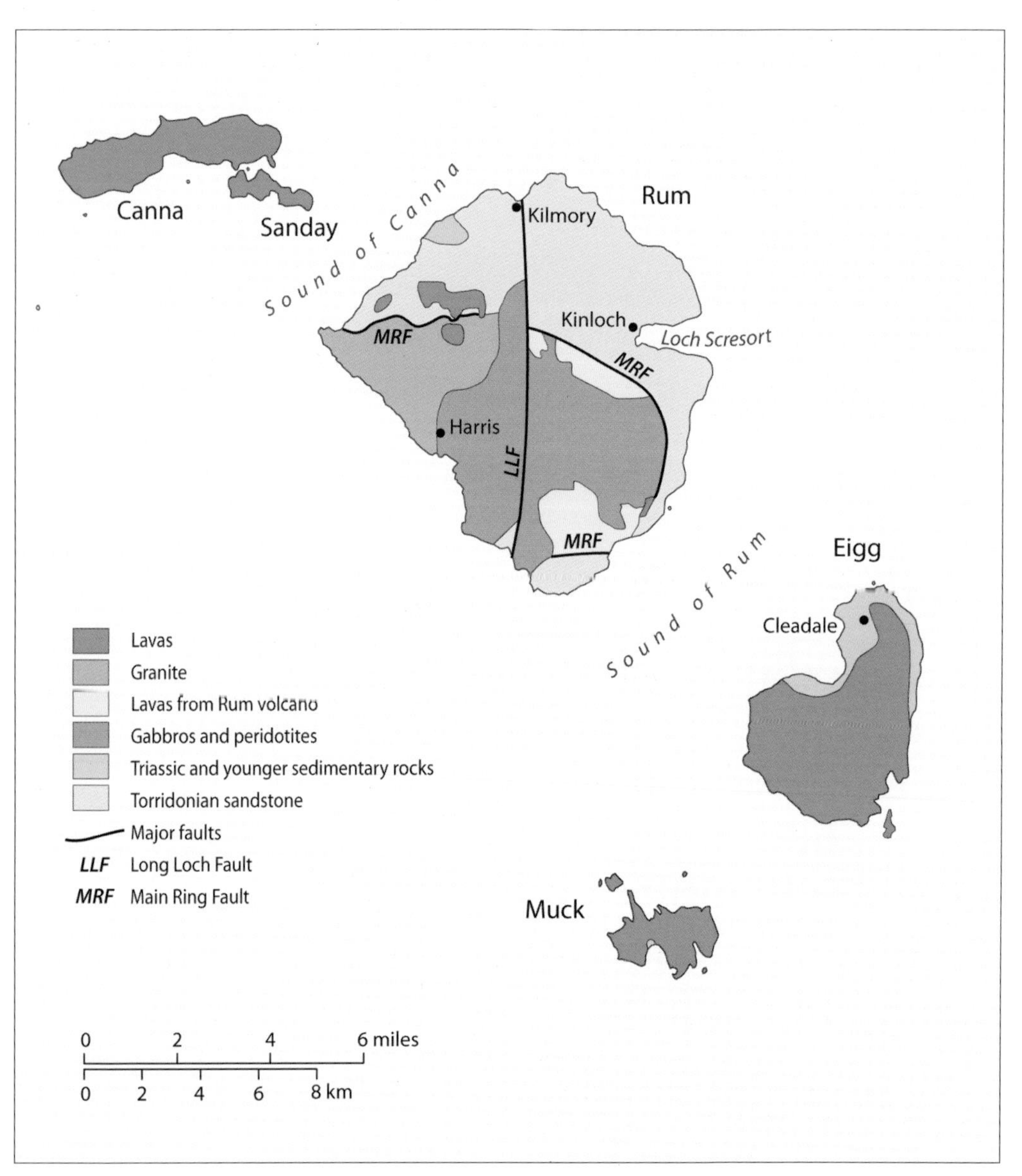

The Small Isles are largely built from the products of three separate volcanoes. The magma chamber of the Rum volcano has been laid bare by erosion, which occurred primarily during the last Ice Age. Glaciers cut deep into the heart of a fiery, albeit now extinct, edifice that last erupted around 58 million years ago. Canna and Sanday are comprised entirely of lavas erupted from the nearby Skye volcano, whereas the lavas that build Eigg and Muck come from a more southerly source – the Mull volcano. On Eigg, Jurassic layers were first described by the fossil adventurer Hugh Miller in his *Cruise of the Betsey*, published after his death in 1857. On Rum, the volcanic rocks are separated from Torridonian sandstones by a fault that has an arcuate trace, known as the Ring Fault. The sandstones are much older, dating back some 900 million years. Tiny patches of Lewisian gneiss, the oldest rocks in Scotland, have also been identified on Rum, but their occurrence is too small to be represented on this map.

1
Time and motion

Time

One of the more perplexing aspects of studying geology is to understand the timescales involved. Geologists talk in terms of rocks being millions and billions of years old, which is alien to most people. Until recent dating techniques gave some clarity on the matter, the age of the Earth was assumed to be around 6,000 years old. A sixteenth-century biblical authority called Archbishop Ussher placed the age of the Earth, with breath-taking precision, as being 'at nightfall on 22 October 4004 BC'. A more scientific method was devised by Professor Arthur Holmes, an Edinburgh-based geologist, in the middle of the last century that allowed more accurate ages to be calculated for many types of commonly found rocks. Towards the end of his illustrious career, he famously said, 'Looking back, it is a slight consolation for the disabilities of growing old to notice the Earth has grown older much more rapidly than I have – from about six or seven thousand years when I was ten, to four or five billion by the time I reached sixty.' The Earth is now known to be 4,543,000,000 – or 4.5 million – years old. But beware, it's at the very heart of scientific enquiry for ideas and numbers to change as new evidence comes to light.

Geologists delight in placing events and evidence of past life in date order. In so doing, we assemble a column or sequence of events with the oldest rocks at the bottom and the youngest at the top. Most is known about the last 550 million years. It was at that point early forms of life became more abundant, diverse and widely distributed across the globe. Many of these primitive creatures had developed hard skeletons that were more easily preserved as the fossil record. This 550-million-year timespan has been divided into more manageable chunks, known as geological 'periods'. Most well-known is the Jurassic Period, as this time and the succeeding Cretaceous were the 'age of the dinosaurs'. Each period has been date-stamped by scientists with a beginning and an end. This allows rocks and fossils locally, nationally and internationally to be correlated and a worldwide picture of geological events to be established.

Pages 8 and 9 are a representation of the geological column for the Small Isles. It tells a story of bursts of frenetic activity at some periods of time, interspersed with long gaps when precisely nothing happened. Or if it did, the evidence for any activity has been removed by later events. This incomplete and fragmentary pageant of geological history for this area is typical of the country as a whole. It's only by assembling a geological column of events for Scotland as a whole that the full geological history of our land can be revealed.

Motion

Another slightly unsettling concept that geological study reveals is that the rocks beneath our feet are constantly on the move. We sit on a thin skin of rock that is being shunted across the globe at the rate our fingernails grow. Slow indeed, but over many millions of years continents can be moved from one side of the planet to another. Collisions between these moving slabs of rock are inevitable and we see the evidence for this in the mountain ranges that are arranged across the globe in liner collision zones. These mountains represent ancient crumple zones, where the layers of sand and mud that accumulated in the intervening oceans and seas were squashed and thrust upwards

Large tectonic plates cover the entirety of the Earth's surface. Earthquakes are frequent at the edges of these plates as they rub and grind past each other. Each plate moves independently, so frequent frictional shocks are inevitable. Volcanic eruptions along these lines of weakness in the crust are another cause of significant instability and seismic activity.

into fold belts as continents collided. The Alps, Himalayas and Scottish Highlands are all places where ancient continents bashed into each other in a never-ending conveyor belt of movement. The unifying concept that explains these collisions and crustal tension is known as plate tectonics. The Earth's crust is comprised of seven massive tectonic plates and around a dozen smaller ones that, taken together, make up the outer skin of Planet Earth.

Movement of the plates across the face of the Earth is powered by heat that radiates from the Earth's core. Despite the age of the planet, the core is still unimaginably hot due to decay of radioactive elements, at 6,000°C or around the same temperature as the surface of the Sun. The hot rocks at depth supply a stream of molten lavas that break the surface, particularly along submarine ridges, lying under the world's oceans, midway between the main continental landmasses. As new material is added, so the continents on either side of the ocean are forced further and further apart. At different locations, the Earth's crust is stretched to breaking point and new oceans form as existing landmasses are pulled apart.

The Earth is truly is a dynamic planet. This is particularly relevant for the Small Isles, as it was in this high-tension environment that the volcanoes that gave rise to Rum, Eigg, Canna and Muck were formed. Much more on this later.

Continents have been on the move for the last three billion years.

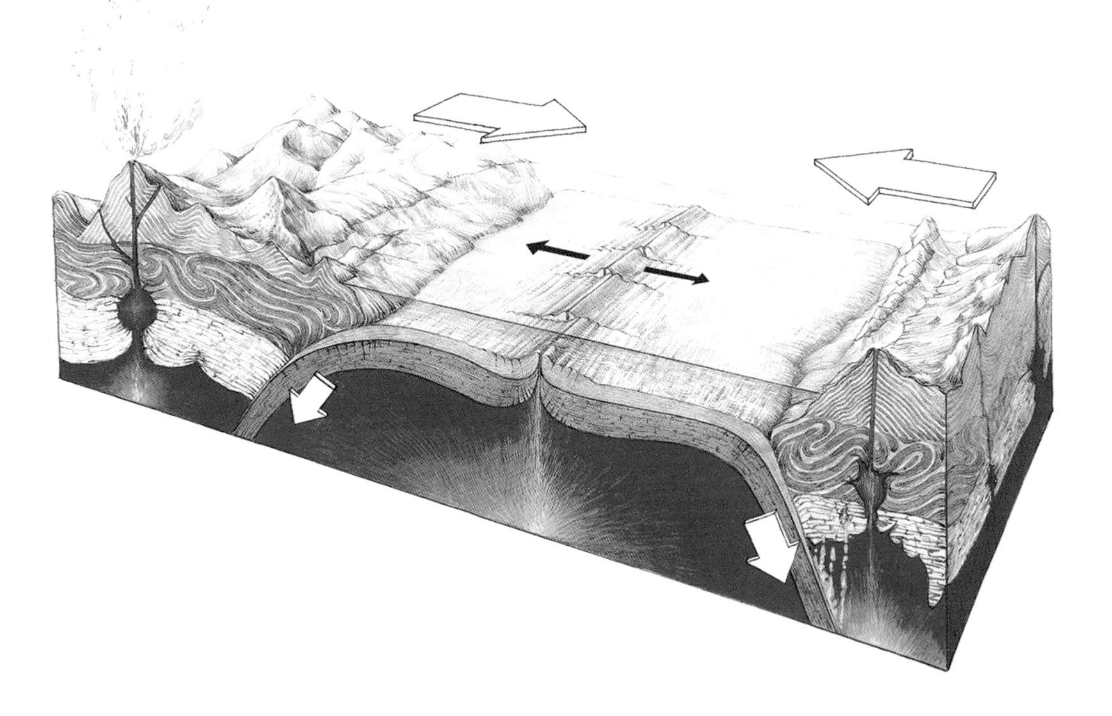

This slice through the Earth's crust gives an idea of how the plate tectonic model works and how continents move. As new material in the form of injections of molten rock is made along the central spine of the ocean, so the continents are forced further apart.

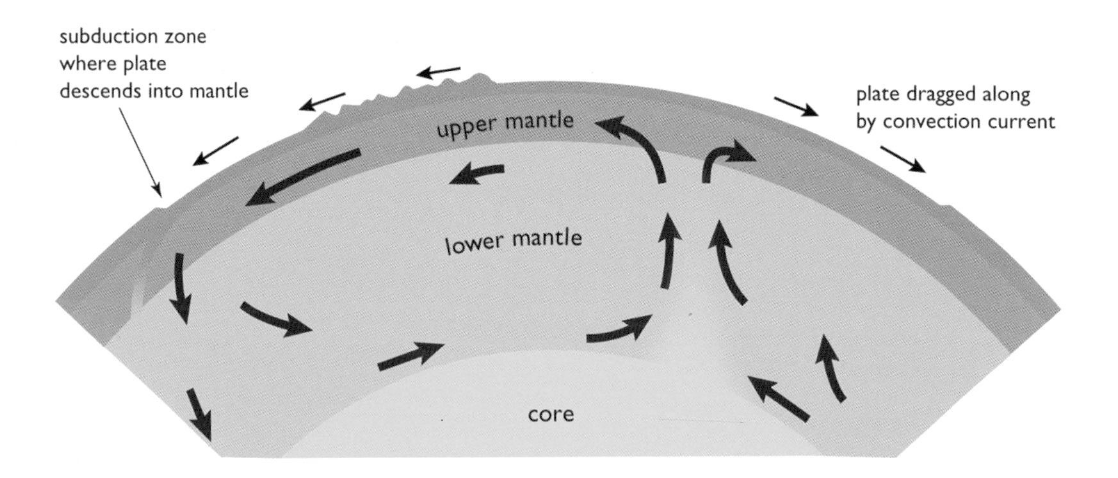

Scotland's journey across the globe was driven by forces deep within the Earth. It was propelled unrelentingly northwards from Torridonian time 900 million years ago through equatorial latitudes to our present position. This journey took our little slice of the Earth's crust through every climate zone the Earth has to offer.

During that time, the land that was to became Scotland moved from a position close to the South Pole to its current location at 57° north of the Equator. During its travels, this chunk of land passed through every climate zone, from the deep freeze of the South Pole, to equatorial temperatures around 290 million years ago during Carboniferous time, when lush rainforests covered the landscape, then onwards to the climate zone we inhabit today. Geology is a story of two journeys: through time, from the earliest crustal rocks formed around 3 million years ago, but also tracking continents across the face of the globe, as they are propelled through the Earth's climatic zones.

2

Ancient times – Lewisian and Torridonian

Lewisian gneiss

The oldest rocks on Rum are known as Lewisian gneiss. They were formed during the very earliest events that happened on the planet and at great depths in the Earth's crust. These rocks have only seen the light of day because of more recent Earth movements that have brought them to the surface. They are intensively folded and faulted, which is testament to their tortuous journey from great depths below ground level. These rocks, although of great interest to geologists, contribute little to the landscape of Rum. Rocks of this age are entirely absent from Eigg, Muck and Canna.

Three billion years ago, when the Lewisian was formed, 'proto' Scotland was unrecognisable from the country we know today. No life existed in this turbulent world. The atmosphere was toxic and the seas were twice as salty as today and devoid of oxygen. Volcanic eruptions were frequent and the early surface of the planet was bombarded from outer space by comets and asteroids. The largest fragments of the Lewisian gneisses are found in the Outer Hebrides and north-west Highlands, with smaller areas forming the bedrock of Coll, Tiree and parts of the islands of Iona and Skye.

Small occurrence of Lewisian gneiss found near Priomh Lochs, near the centre of Rum. The outcrop has been smoothed by the ice in recent geological times, enhancing the banding which is so characteristic of rocks of this age.

Torridonian sandstone

In date order, the sandstones that build the northern and eastern quadrant of Rum are the next chapter of the Small Isles to be described. These rocks, known as Torridonian sandstones, form a stepped landscape of sandstone layers deposited by rivers that flowed over an ancient landscape. These terraces were originally laid down as a series of flat layers, but were later tilted to the west by between 10° and 30° by subsequent Earth movements. Each layer represents a separate flood event, when the seasonal river burst its banks and flowed across the surrounding floodplain, leaving a burden of sand and mud on the surface. The deep red colouration of the sediments comes from the fact that the iron compounds that bind the sand grains together have been oxidised through exposure to the air. Iron oxide is red in colour. Some of the Torridonian layers have a higher content of mud flakes, indicating they were laid down at the bottom of a lake which subsequently dried up and the mud fragments were ripped up by fast-flowing rivers.

Some of the hills on the mainland are made of similar rock, which were formed at the same time, notably Slioch, Suilven and the valley sides of Glen Torridon, from which these rocks take their name. This thick carpet of sands and gravels completely buried the ancient Lewisian gneisses to a variable thickness, of around two kilometres.

The Lewisian and Torridonian rocks were formed when 'proto' Scotland was close to the South Pole! Amazing, but true. What followed, over the succeeding 650 million years, was a slow northward drift, driven by the movement of the tectonic plates.

Torridonian sandstones underlie the northern area of Rum. Despite the incredible age of these rocks at around 900 million years, they are remarkably untouched by later alteration. Some layers contain pebbles, eroded from the bedrock over which the ancient and now long-disappeared rivers flowed. The oldest strata, at the bottom of the sedimentary pile, contain lumps of locally derived Lewisian gneiss.

3
Mesozoic rocks – Triassic and Jurassic

Triassic sandstones on Rum

We pick up the story of the rocks that built the Small Isles during Triassic times. Scotland had migrated north of the Equator and was part of a much larger landmass known as Pangaea, from the Greek meaning 'All Earth'. Land stretched from pole to pole. Scotland was landlocked, joined to the south and west to the Americas, and to the north to Europe. It's a crazy map of the world that existed during Triassic times, but that is the nature of how plate tectonics affects the ever-changing geography of our planet.

Sedimentary layers that date from Triassic times are found to directly overlie the Torridonian sandstones in the north-west of Rum on Monadh Dubh. It's amazing to think that the junction that separates these two deposits represents around 650 million years of

This is how the distribution of land and sea looked during Triassic times (around 250 million years ago). Scotland lay just to the north of the Equator and was part of a large area of desert.

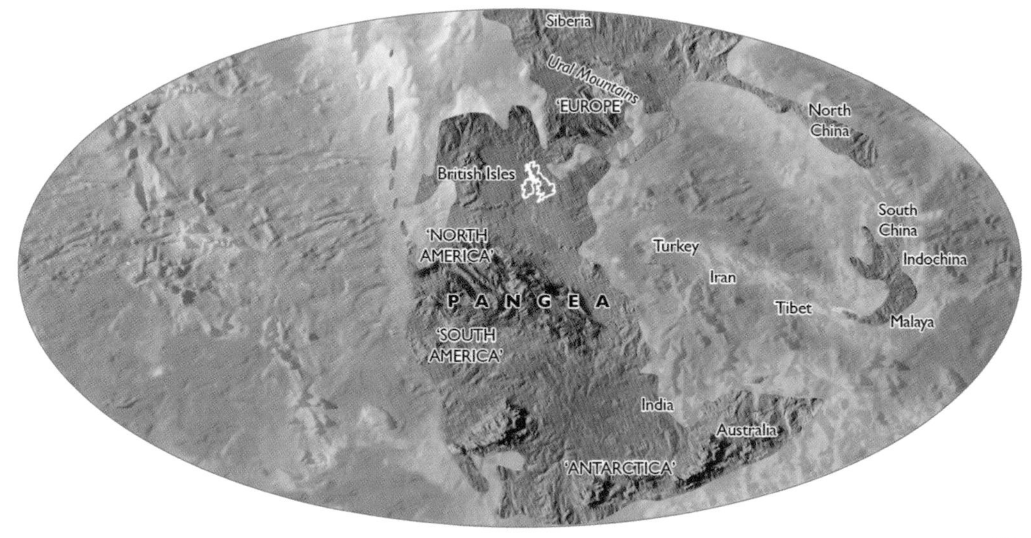

These sedimentary layers of Triassic age sit atop the Torridonian rocks north of Glen Shellesder on the north-west coast of Rum. The white layer is the caliche horizon, which is indicative of a semi-arid environment of deposition.

geological time. Also Scotland had moved an astonishing distance between the time of the Torridonian and the date the overlying Triassic rocks were laid down. It's difficult to compute exactly, but it's in the order of around 10,000 kilometres, from the South Pole to just north of the Equator. It's these mind-blowing facts that make geology such an amazing subject to study.

The Triassic deposits are indicative of a desert environment. Thin layers of caliche or cornstones are key to this interpretation. Today, these layers of hardened calcium carbonate form in deserts and other semi-arid environments, like the Kalahari Desert, so we can assume these rocks on Rum formed under similar conditions some 250 million years ago. Minute crustaceans called ostracods and poorly preserved plant remains have also been recovered from these rocks. We shouldn't be surprised by the interpretation that these rocks were laid down in such an extreme environment, as this area was positioned around the latitude of the Sahara Desert at the time.

Jurassic rocks of Eigg and Muck

Our focus changes from Rum to Eigg and Muck for the next step through geological time. The rocks of the northern area of Eigg and a small patch on the south coast of Muck date from Jurassic times. Their initial study is the stuff of legend. Hugh Miller (1802–56) is one of Scotland's best-known early geological pioneers. As a young man, he studied the fossils of his native Cromarty on the east coast and maintained a lifelong interest in the remains of ancient life he found entombed in the rocks. In the summer of 1845, he undertook a cruise around the islands of the west coast of Scotland that he later immortalised in his book *The Cruise of the Betsey . . . with the rambles of a geologist.* He made remarkable discoveries on his flying visit to Eigg. Miller found the bones of a plesiosaur, a long-extinct marine reptile, the first to be described from Scotland. The equally renowned fossil collector Mary Anning (1799–1845) had already unearthed specimens of a similar type from the cliffs of Lyme Regis in Dorset in the 1820s, but this was the first find of its type in Scotland. It immediately identified these rocks as being of Jurassic age.

Since that early excursion into the field by Miller, hugely exciting fossil finds have been made at the Hugh Miller Bone Bed on Eigg. The site has been re-investigated several times since Miller's day. The bones of marine turtles, crocodiles and shelly fossils such as gastropods and bivalves, along with fish scales and plesiosaur remains, including vertebrae, ribs, teeth and disarticulated skull bones, have all been recovered from these rocks.

Most impressive of all is a serendipitous find made in 2020. Researcher Dr Elsa Panciroli was running along the beach, catching up with colleagues, when she spotted a bone embedded in a boulder. Closer inspection revealed that it was the 50-centimetre-long hind leg of a stegosaurus dinosaur. In Scotland, discoveries of dinosaur bones have only previously been made on the Isle of Skye, Scotland's Jurassic Park, so this discovery is a scientific game-changer.

The bone and surrounding deposits have been dated at 166 million years, from Middle Jurassic times. Dinosaur bones from this time are rarely found anywhere in the world, adding greater significance to the discovery. Stegosaurus dinosaurs are heavily built, plant-eating animals, with large plates on their backs probably to regulate their temperature. Spikes on their tail were largely for defending themselves from predators.

Further evidence that stegosaurus roamed Jurassic Scotland is found in dinosaur footprints discovered in Skye, which were probably made by the same species of animal. The bone has been damaged by the pounding of the waves since it was partially released from the rock that encased it. There is also evidence that it was nibbled by scavengers after the dinosaur's death. The bone is now in the care of National Museums Scotland (NMS) in Edinburgh.

The Jurassic rocks of Eigg also provide abundant evidence about the conditions that prevailed in this part of Scotland 170 million years

This reconstruction was drawn by Dr Panciroli to illustrate what the stegosaurus would have looked like in life. Individuals were known to be up to almost 5 m in height and weigh up to 2.5 metric tons. It would have been an extraordinary sight to see these behemoths trundling across the plain 166 million years ago. Most people will have seen *Jurassic Park* – the movie. We had our own version of it right here on the land that would become the Small Isles!

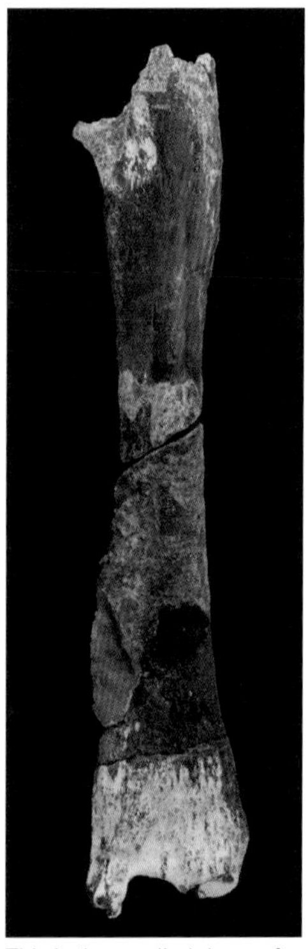

This is the rear limb bone of the stegosaurus dinosaur. It has been extracted from the rock, carefully prepared and is now part of the NMS collection.

ago. The other fossils recovered from these rocks, including bivalves and gastropods, and the nature of the strata themselves, gave rise to the name 'The Great Estuarine Series' (now styled 'Group') given by the early geological surveyors. It is thought that these rocks were laid down in an extensive muddy estuary that extended, in Jurassic times, from Muck to Skye. Lying on top, albeit still part of the Great Estuarine Group, is a thick layer of sandstones that indicate a change of environment. These sandstone layers are evidence for a series of deltas that developed where ancient and now long-disappeared rivers met the sea. These rivers flowed westwards from the Scottish mainland, as it existed at that time.

Sedimentary layers also occur at Camas Mor on the south coast of Muck and date from this time. These strata are fossil-bearing, which allows their correlation with rocks of a similar age on Eigg and Skye. The extent of the great estuary that existed some 170 million years ago in this area can be appreciated by joining these now disparate patches

Embedded in the delta sandstones are these amazing structures called concretions. This example formed shortly after the host rocks (the delta sandstones) were deposited. Fluids circulating through the soft sedimentary layers, in this case calcium carbonate-rich, derived from mollusc shells, coalesce round a nucleus, often a small fossil, and layers of the harder material cemented together to form a ball. They weather out from the host rock, as the calcium carbonate-rich cement forming the concretion is harder than the host sandstone.

of strata. They would, at one time, have been an unbroken sequence of rocks deposited in this extensive but now completely vanished estuary environment.

4

A tale of three volcanoes

We have already heard about the mechanism that can shift continents and create mountains: plate tectonics. The distribution of land and sea is ephemeral and ever-changing on a geological timescale. The geography of the Earth changes as the underlying currents in the mantle shunt continents around the surface of the planet. Continents combine as they collide and, later, they can also split apart. Around 65 million years ago, Pangaea, the supercontinent of which Scotland had been part, started to disintegrate. A great crack developed that split Europe from North America, and as a result of this latest con-

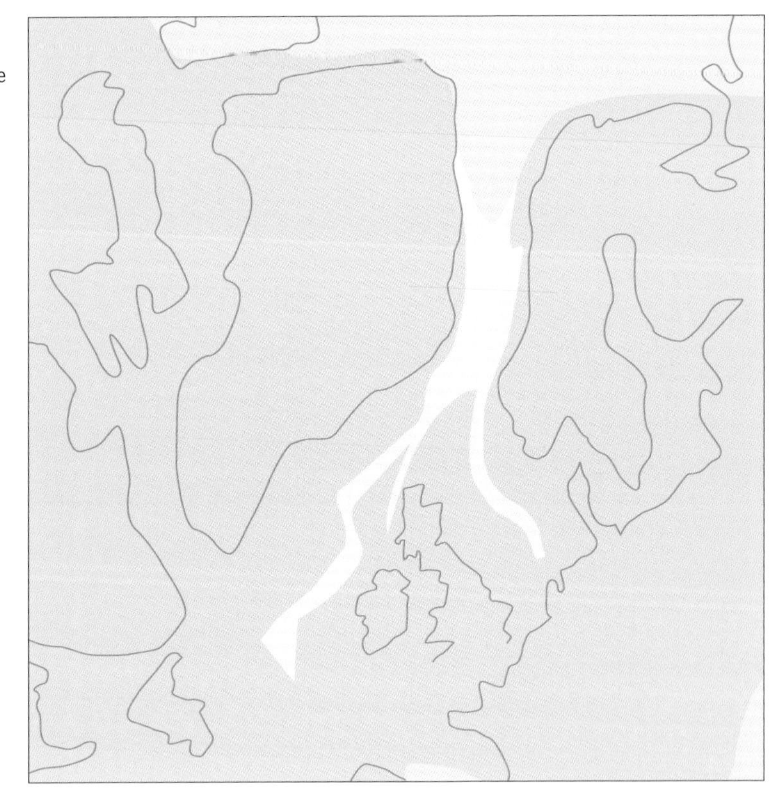

As Pangaea began to break up, Europe and North America parted company. The crust thinned to breaking point, as these continental plates pulled further and further apart.

tinental rearrangement the North Atlantic Ocean was created. A new world order was fashioned from these building blocks hewn from the ruins of Pangaea.

The Small Isles owe their origin to these cataclysmic global events. As tensions built in the crust just to the west of Scotland, as it existed 65 million years ago, explosive events were unleashed. A series of volcanoes reared up from the widening ocean floor, and over a period of many millions of years spewed lava across the ancient landscape and hurled ash high into the air. Lava fields extended south-west to the Rockall Plateau and north-east beyond the Faroe Islands. The volcanic activity related to the break-up of Pangaea was explosive, extensive and extremely long lived. Puncturing the thick layers of lava between Scotland and Iceland is a series of around 30 smaller volcanic centres, known as seamounts. These mini-volcanoes don't break the surface of the Atlantic Ocean, but they provide further evidence of the intense volcanic activity that accompanied the break-up of Pangaea.

Closer to home, along a zone that we now recognise as lying just to the west of the Northwest Highlands, the crust was at breaking point. Large volcanic centres burst into life. St Kilda (the youngest of them all), Skye, Rum, Ardnamurchan, Mull, Arran and also Ailsa Craig were all fire-breathing monsters that lit up the ancient skies for many millions of years. Rock-dating techniques allow us to give an age for individual eruption events with a degree of accuracy. The Skye volcano was the longest lived, with volcanic episodes taking place over a prolonged period of just under 6 million years.

The Earth's crust cracked as a result of the tensions created as the continents of Europe and North America continued to move apart. This allowed molten magma to well up from below and, periodically, to erupt in violent and sustained volcanic events. Huge reservoirs of magma accumulated just below the surface and fed the periodic eruptions. Where the magma cooled without being erupted, these rocks provide invaluable information about the inner workings of an active volcano. The prominent peaks of Askival and Hallival on Rum are arguably the best example of a 'frozen' magma chamber in the British Isles.

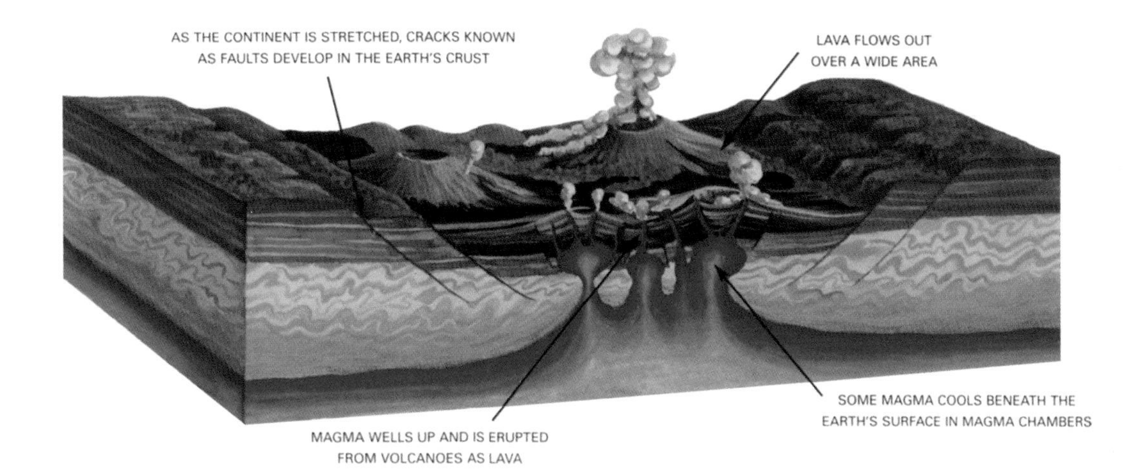

This map shows the origins of the lavas that built the Small Isles. The Eigg and Muck lavas originated from the Mull volcano to the south. The Skye volcano's associated lava field extended southwards and built Canna and Sanday, and also contributed small patches of lava on Rum.

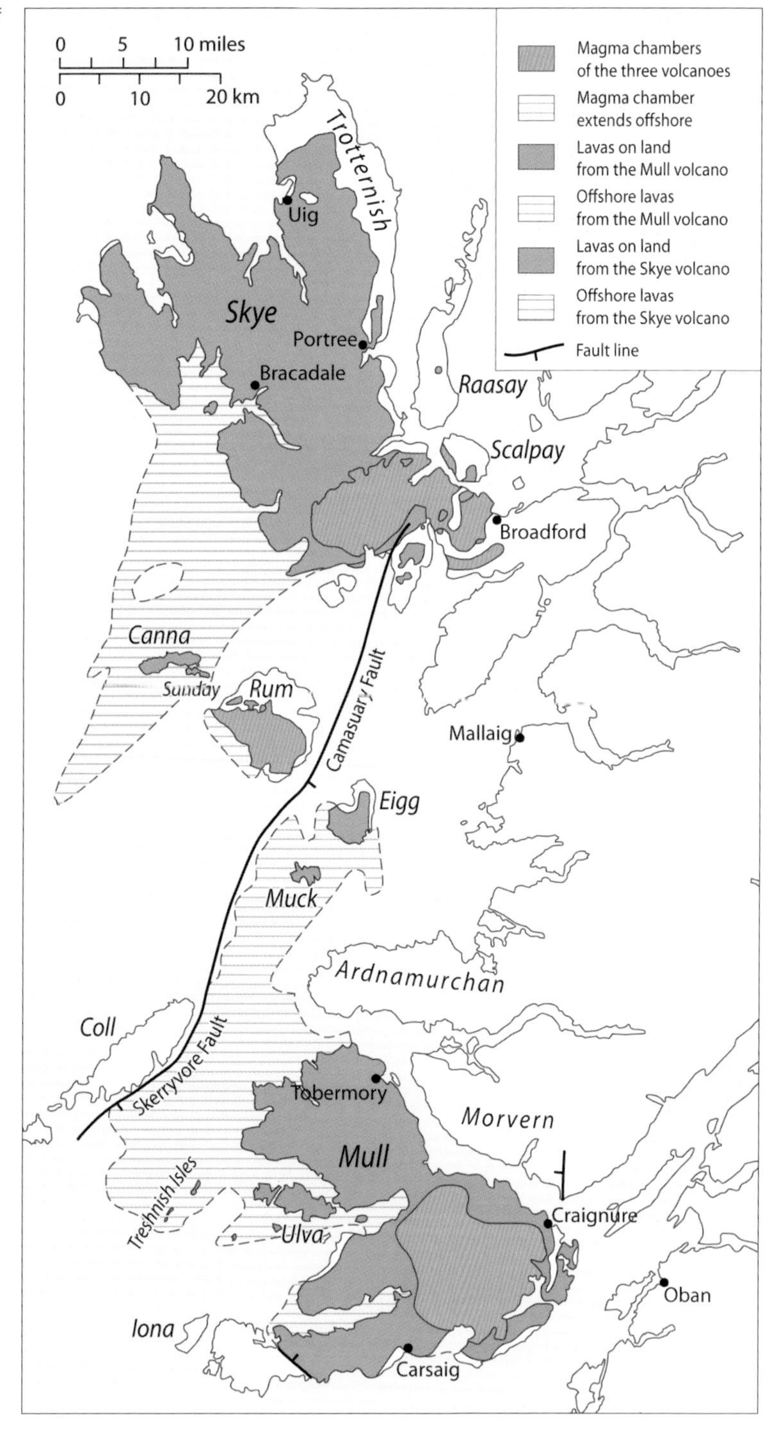

Of the four Small Isles, only Rum was a volcano. The other three islands are built from lavas that flowed from the neighbouring eruption sites of Skye and Mull. So the geological history of the Small Isles is the tale of three volcanoes: Rum, Skye and Mull.

The Rum volcano

Rum has the most complex story to tell, so we'll start there. All that remains of the Rum volcano is the magma chamber that was once located deep underground. It now sees the light of day, as the volcano's upper structure has been completely stripped away by the action of later erosion. Study of these rocks from the mid nineteenth century onwards by some of the most learned pioneers of our science has revealed a very detailed story of how the volcano functioned.

The initial phase of volcanic activity involved huge quantities of molten rock rising from the depths of the Earth to create a great pool of magma a few kilometres beneath the surface. This caused the land, as it existed at that time, to rise into an unstable domed structure, like a boil on the skin waiting to burst. The Torridonian sandstones lying above the volcano, which had been undisturbed for around 900

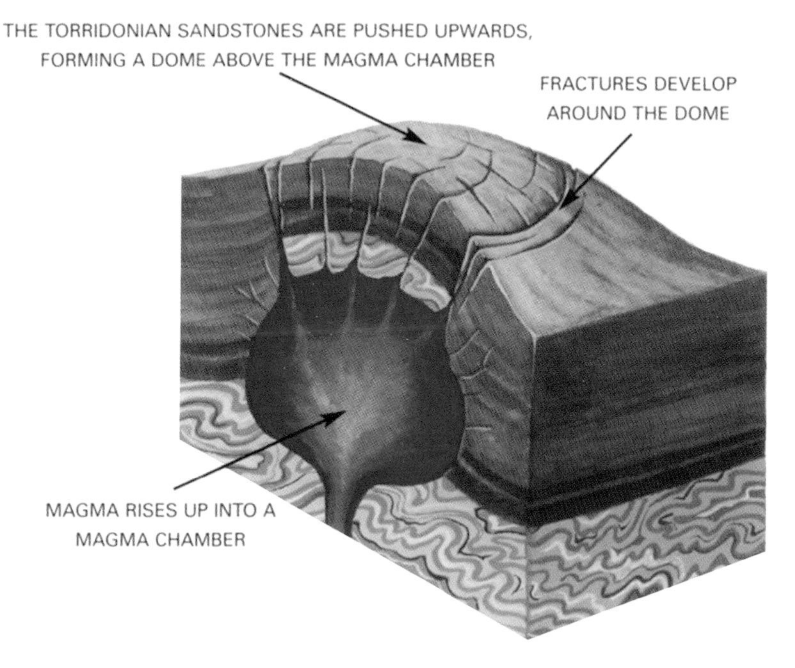

Molten rock rose from below, as the Earth's crust was stretched in response to tectonic plate movement. The surface was jacked up, as a result of the influx of molten rock, creating a radial pattern of fractures in the overlying rocks.

The edifice tumbled down to form a cauldera, as the dome could no longer support its own weight. Great clouds of searing hot gas and ash, known as a pyroclastic flow, were released to the surface, which then billowed across the surrounding landscape. A modern day analogy is the ash cloud that engulfed Pompeii in AD 79. The main difference is that this eruption took place 60 million years before any human population existed.

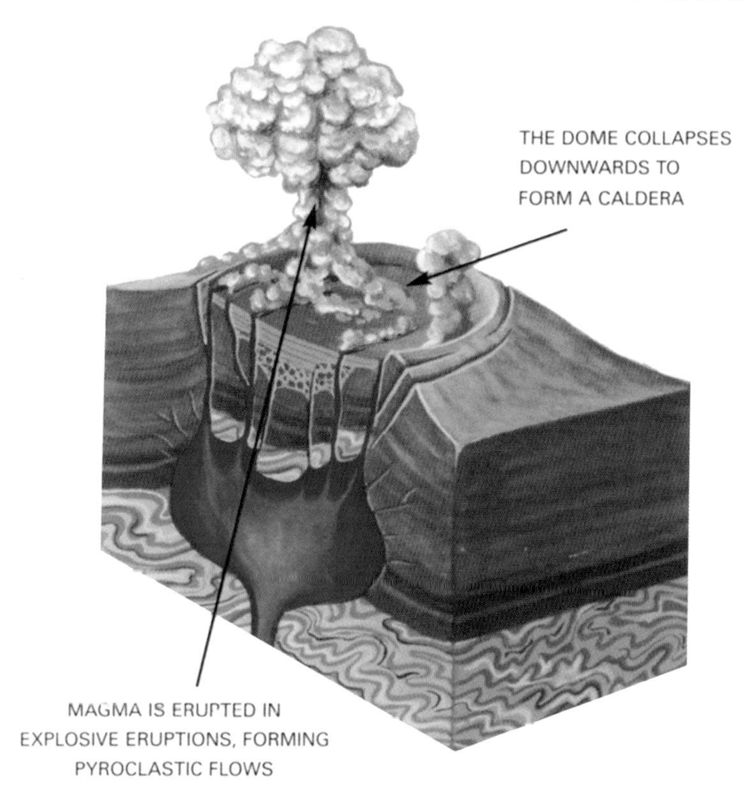

million years, cracked and fractured in response to this intense pressure from below.

The dome rose to more than a kilometre in height. A reasonable estimate can be made of the height of this structure because of the manner in which the adjacent sandstones have been bent and buckled from their original horizontal position. Inevitably, this unstable structure eventually collapsed in on itself, forming a chaotic jumble of rocks and molten magma. The fracture pattern of the overlying rock was roughly circular, so what appeared at the surface was a crater-like structure that geologists call a cauldera. Rocks cascaded down the sides of the rapidly deepening hole, with landslides and rockfalls adding to the accumulating debris. This jumble of magma, ash and fallen blocks is recognised today in Coire Dubh.

The composition of the magma at that stage of the volcano's development was rich in quartz and feldspar – two of the key component minerals that make up granite. This molten mix cooled slowly to form the hills that build the western part of the island. Their extent is clearly

These exposures in Coire Dubh, in the central area of Rum, are the remains of the pile of debris that accumulated as the cauldera developed. The blocks are largely comprised of Torridonian sandstone which collapsed into the fiery pit below. This collection of angular and partly rounded blocks of sandstone and occasional lumps of Lewisian gneiss is known as a breccia.

defined on the geological map on page 10. The geology of this part of the island is described as the Western Granite, with the peaks of Orval and Ard Nev being the most well-known.

Later pulses of molten rock to reach the still-active magma chamber of the Rum volcano had a different chemistry. Described by geologists as base rich, this melt had a higher content of magnesium and iron. When it cooled to form a solid mass, this melt would be recognised as a gabbro, or a variety of other related rock types with picturesque names such as feldspathic peridotite or bytownite troctolite. The Black Cuillin on Skye is made from similar material. Evidence that these two distinct magma types came into contact is provided by

Above.
Two main magma types – pink granite and dark gabbros – fuelled the Rum volcano. There was clearly close contact before they cooled and solidified to form a maze of broken and interconnected veins, and blocks of pink and black rock.

Right.
The magma chamber under the Rum volcano was supplied with molten rock from deep in the Earth's crust. As the melt cooled, crystals formed and fell to the floor of the chamber, building up in rhythmical layers. Askival and Hallival show this layering process to perfection.

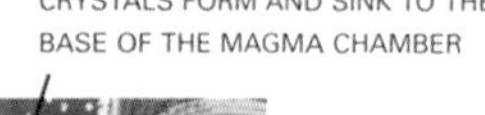

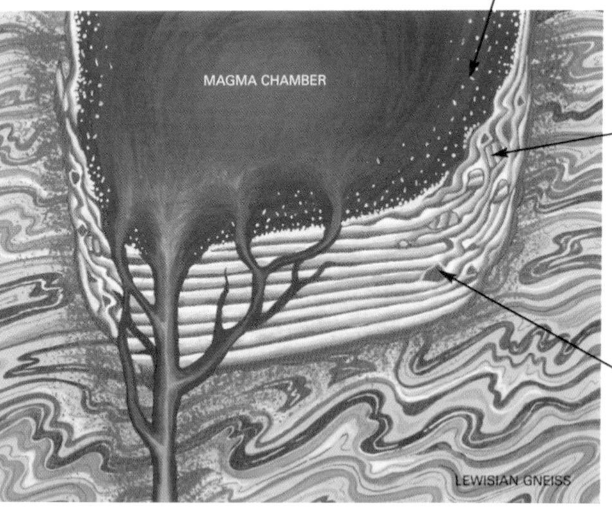

this spectacular exposure on the south side of Harris Bay in the far west of the island.

From this point on, magma that gave rise to base-rich black gabbro made up the greater part of the molten rock that found its way to the magma chamber underlying the Rum volcano. What has attracted the attention of geologists over the last century is the fact that these gabbros accumulated in layers on the floor of the magma chamber. Askival and Hallival exhibit particularly fine examples of this phenomenon that stands favourable comparison with similar layered sequences from Greenland and South Africa.

Some of the magma may have been erupted from the Rum volcano as basalt lava, but none remains today. Lavas that flowed from the neighbouring volcano of Skye are perched high on Bloodstone Hill. After careful study of their mineral content and chemical signature, they can be matched to the Skye volcanic centre with a high degree of certainty.

The mountains of Hallival (illustrated here) and the adjacent Askival are the most prominent peaks on Rum. They both show this layering that resulted from the settling of crystals on the floor of the magma chamber

Layering of the gabbros is equally obvious at ground level.

Bloodstone Hill in the far west of Rum is capped by basalt lavas that flowed south from the Skye volcano. They are just small fragments of a more complete blanket of lavas that would have formerly existed. The coverage was dissected during the Ice Age when the ice cut deep into the surface layers.

Isle of Eigg

The Sgurr of Eigg is one of Scotland's most iconic views. A bare, steep-sided ridge runs almost the full width of the southern end of the island. It is largely made from pitchstone, an unusual volcanic rock that takes the form of glass.

The pitchstone formed into a series of columns as it cooled. This volcanic layer sits on top of a thick layer of conglomerate that contains fragments of pine wood. To complete the picture, the Sgurr was erupted over older lavas that form part of the lava field associated with the Mull volcano. This complex sequence of events has been studied

This is the Sgurr of Eigg, showing the upper layers of pitchstone lava formed into vertical pipes.

Above.
Basalt columns formed in the lavas that built most of Eigg. The Sgurr of Eigg lies beyond.

Opposite.
An aerial view of Eigg shows a stepped landscape, where each tread represents, from right to left in the picture, a different, and younger, lava flow. The flows are of variable thickness and would have not necessarily been erupted at regular intervals of time. It demonstrates the long-lived nature of the volcanic activity that emanated from the Mull volcano to the south.

and described in print by Victorian, Edwardian and later geologists of impeccable repute, including Hugh Miller, Sir Archibald Geikie, Dr Alfred Harker and Sir Edward B. Bailey. They offered a variety of different explanations for this curious arrangement of rock.

The modern interpretation is that the pitchstone is not actually a lava at all but the cooled remains of a pyroclastic flow of white-hot ash and gas. A valley had been cut into the basement rocks along which a river had flowed, depositing boulders and incorporating the remains of pine trees that flourished on the valley sides. The pyroclastic flow followed the same course. Later erosion removed much of the adjacent bedrock but left the pitchstone standing, as it proved to be much more resistant to the ice, wind and water that moulded the contours of the landscape into the familiar shapes we see today.

A final footnote on the Sgurr: it is thought that the eruption of the pitchstone is the final event recorded in the Small Isles volcanic story. In this area, studded with volcanoes, this event represented a highly explosive eruption and the last gasp of a sequence of dramatic episodes that left an indelible mark on the bedrock of the area.

Canna and Sanday

Another vital aspect of the geological story of the Small Isles is revealed on Canna and Sanday. The narrative so far conjures up a vision of constantly erupting volcanoes, restless magma chambers and pyroclastic ash flows. But there were many tranquil interludes between these dramatic episodes. The lava flows that built Canna and Sanday originated from the Skye volcano to the north. When the eruptions were silent, great rivers flowed across the barren landscape. The evidence for their existence is provided by thick sequences of conglomerate. These deposits are choked with large, rounded boulders, smaller

rocks and sand lenses that interleave between successive lava flows. Some of these sedimentary layers reached up to 80 m in thickness.

The nature of the boulders and pebbles help us to reconstruct the direction of flow of the river system. Lumps of Lewisian gneiss and Torridonian sandstone predominate in these deposits, which would suggest a direction of river flow from the north and east. Deposits of agglomerate, which comprise layers of ash mixed with fragmented lava, have also been recorded as an integral part of the layers of sediment. So even during times of tranquillity, there were occasional eruptions of ash and lava. Archibald Geikie, an early geological pioneer who seems to have been ubiquitous in his travels across Scotland, published a geological account in 1897.

Opposite.
These coastal cliffs on Canna, and outcrops on Compass Hill, expose thick conglomeratic layers comprising rounded boulders of locally derived lavas, occasional lumps of Lewisian gneiss, Torridonian sandstone and granites identical in nature to those on Rum. They are sandwiched between thick lava flows. Some of the lumps are around 2 metres in size, which gives an indication of the strength of the current and therefore the size of the rivers that flowed across the landscape between volcanic eruptions.

Isle of Muck

Muck and Eigg share the same source of molten lava: the Mull volcano. The bedrock of Muck is a variation on the theme described from the other Small Isles: namely lava, pyroclastic flows and layers of sediment deposited between the eruptions. Thin pulses of basalt rock, called dykes, have sliced through the bedrock to create near-vertical structures. These features have an average thickness of around 2 metres and can be traced for long distances across the Small Isles and also mainland Scotland. Each volcano that was active at this time pumped out a north-west–south-east trending swarm of these dykes. One of these dykes on Muck is particularly thick and it has thermally altered the rocks into which it came into contact when still molten. These thermally altered rocks are known as hornfelses. An interesting assemblage of calcium-rich minerals, indicative of high temperature alteration, has been described from rocks adjacent to this dyke on Muck. The force required to inject these volumes of magma, in some instances hundreds of kilometres through pre-existing rock masses, is colossal and difficult to comprehend.

5
The Ice Age

Opposite top.
The erosive power of the ice is clearly demonstrated by observing the valleys cut in the bedrock by these Alpine glaciers.

Opposite bottom.
The jagged peaks of Askival (foreground), with Trollavall on the right and Ainshval beyond, were shaped by the action of the ice. Great gouges were scooped out of the bedrock and valleys were cut by the erosive force of the ice. Frozen water the ice might be, but, on the move, it cuts like a saw into the toughest bedrock. The lower surface of the ice, in contact with the bedrock, bristles with stones, so as the glacier moved, an irresistible pressure caused erosion of the rock surface below.

Our changing climate is not just a modern phenomenon. It's been a factor throughout geological time. In the last 2.4 million years, much of northern Europe was plunged into the deep freeze of an Ice Age. The temperatures weren't constantly below zero, as more balmy conditions, similar to or hotter than today, did punctuate the longer periods of Arctic conditions. These are called inter-glacial periods. We're experiencing such a warmer interlude at the present time and have done so since the last ice sheet melted some 11,000 years ago. The ice will come back at some point in the future and a blanket of ice and snow will, once again, extend from the North Pole.

An ice sheet of variable thickness covered Scotland for over 2 million years. It extended south from the polar region, and waxed and waned in thickness and extent in response to changing ambient temperatures. An abundance of evidence tells us that the land was held in an icy thrall for extended periods and the landscape was significantly altered as a result. Glaciers inched their way from the higher ground towards the sea and, in the process, left an indelible mark on the countryside. The higher peaks were shaped into a softer, more rounded profile by the passage of the ice, whilst in other locations deep gouges were carved into the bedrock.

The deep freeze reached its most severe around 20,000 years ago, when most of the British Isles were encased in snow and ice. The greatest thickness of ice was over Rannoch Moor at around 1,000 m thick, with the blanket of ice and snow tapering in depth eastwards and westwards from that point.

As the climate warmed up at the end of the last advance of the ice, a great thickness of boulders, stones, sand and mud, picked up by the ice as it moved across the land, was dumped in a series of chaotic, albeit characteristic, mounds and ridges. Many will remember from school the names of these landscape features: kames, eskers and drumlins. Corries – the deep circular pits created by the erosive force of the ice – will be an equally familiar term.

This smoothed rock surface is another example of the work of the ice. Its shape demonstrates that the ice moved from east to west, which would fit with the larger picture of ice moving laterally from the area of thickest accumulation, the Scottish mainland.

The Small Isles didn't escape these freezing conditions, and the hills and glens we see today bear all the hallmarks of a heavily glaciated landscape.

6
Human habitation and land management

Archaeological finds

The end of the current phase of the ice age came around 11,000 years ago, when the climate warmed rapidly with the onset of more benign inter-glacial conditions. The land that became the British Isles was a small promontory on the northern margin of the emerging continent of Europe. Early nomadic tribes explored the newly ice-free landscape and some, it appears, found their way to this part of the world. Sea levels worldwide were lower, so these early explorers could have walked from what is now mainland Scotland to the Small Isles.

Arrowheads, scrapers and rudimentary blades have been recovered from the soil along what is now the shore of Loch Scresort on Rum. These precious artefacts have been dated at around 9,000 years old, which places them as perhaps the earliest signs of human habitation in Scotland. These early craftspeople used a type of rock which was in plentiful supply on Rum: the agates of Bloodstone Hill. They are similar in their properties to flint and can be worked to create razor-sharp edges that can slice meat and scrape hides. The archaeological record also reveals signs of human settlements in the sheltered glens of Rum, and later cairns and Iron Age forts at Kilmory.

On Eigg, Iron Age hut circles are frequent, dating back to around 800 BC. And on neighbouring Muck a Bronze Age dagger has been recovered, as well as built structures dating from the Iron Age.

Excavations on Canna have revealed evidence, in the form of shards of pottery, of human occupation during Neolithic or New Stone Age times. And later, huts and field walls date from the Bronze Age.

MODERN OCCUPATION AND MANAGEMENT

Rum

Perhaps the most arresting architectural sight on the Small Isles is Kinloch Castle on Rum. It was the pleasure palace of the Bullough family, Lancastrian textile barons who bought the island in 1888. Kinloch Castle was their Highland retreat, which they built of red sandstone, imported from the Isle of Arran.

Rum was sold to the Nature Conservancy in 1957 and the whole island has been managed as a National Nature Reserve since that time. There is spectacular bird life on the island. Golden eagles and white-tailed eagles have made this place their home, as well as a mountaintop colony of Manx shearwaters, which comprise around 40 per cent of

This rather incongruous structure is located near the head of Loch Scresort on Rum.

the British population of this species. Guillemots, kittiwakes and red-throated divers also contribute to this stellar list for birders to observe. Research on the red deer population on Rum has been conducted since 1953, allowing data to be collected on animal behaviour, population dynamics, natural selection and the effects of weather. A wide variety of habitats and plant and bryophyte assemblages are also of interest, as is the varied insect life.

The majestic sea eagle in full flight having taken a fish from the water.

Manx shearwater.

The Island of Eigg

Eigg was owned by a succession of lairds and entrepreneurs until a buy-out by the Isle of Eigg Community Trust was negotiated in 1997, with the help of government aid and private donations. The residents set about making good many deficiencies that had plagued the islanders, particularly lack of infrastructure.

According to the Isle of Eigg website, within ten years of the buy-out, five properties had been renovated on the island, including a new multi-purpose centre, a shop, a post office and a tearoom, while a craft shop had also been constructed. Other upgrades included an island broadband network and a renewable electricity grid. The population of the island is also on the increase, with young people setting up home on Eigg and starting businesses.

Tourist facilities on the Isle of Eigg.

Canna and Sanday

Canna is in the care of the National Trust for Scotland (NTS). It was gifted to the Trust in 1981 by its then owner John Lorne Campbell and his wife, Margaret Fay Shaw. It is still run as a farm. Efforts have been made to retain residents and attract new ones – a responsibility that has now been devolved to the local community.

The Isle of Muck

The island is still in private ownership, its pastures are largely used for grazing. The main occupations are farming, tourism and the making of rugs and woollen garments.

Port Mor harbour on the Isle of Muck.

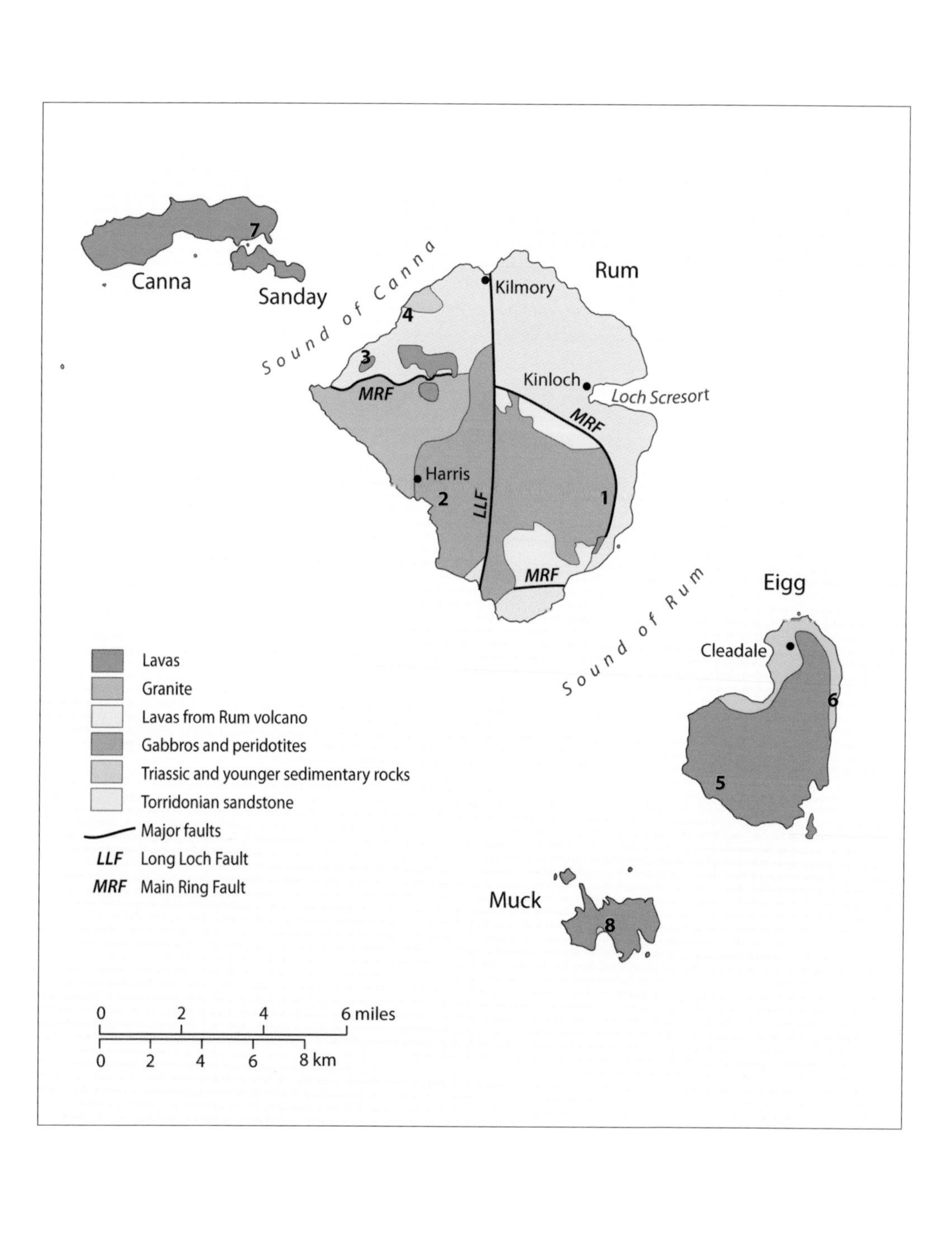

Canna
Sanday
7
Sound of Canna
Rum
Kilmory
4
3
MRF
Kinloch
Loch Scresort
MRF
Harris
2
LLF
1
MRF
Sound of Rum
Eigg
Cleadale
6
5
Muck
8
Lavas
Granite
Lavas from Rum volcano
Gabbros and peridotites
Triassic and younger sedimentary rocks
Torridonian sandstone
Major faults
LLF Long Loch Fault
MRF Main Ring Fault
0 2 4 6 miles
0 2 4 6 8 km

7

Places to visit

This list of key sites should help those who are unfamiliar with the area find some interesting places to visit. It's not an exhaustive list, but it's a start. Two excellent geological field guides cover Rum and Eigg respectively, but they are really for amateur and professional geologists, with a good knowledge of the subject. Both are available from the Edinburgh Geological Society's website. The Small Isles are covered by the 1:25,000 scale OS Explorer series Sheet 397 and the Bedrock Geology UK North map, published by the British Geological Survey. These maps will help you plan and execute your visit to the area. These sites are all described in greater detail in the foregoing text.

Rum

1. **Askival and Hallival:** the most prominent peaks in the south-east quadrant of Rum. They are made from the solidified remains of the magma chamber that lay beneath the Rum volcano. They are a feature of international geological importance and well worth a visit.

Eigg (left), with the mountainous terrain of the south end of Rum, including Askival and Hallival.

2. Harris Bay: on the south-east side, the exposure described and illustrated on page 29 can be found. It shows the two main types of magma associated with the Rum volcano: the pale 'acidic' granite and the black base-rich gabbro in intimate contact.

3. Bloodstone Hill: illustrated on page 30, this is built from Torridonian sandstone and topped by lava flows from the Skye volcano. The hill takes its name from the small inclusions of agate in the lavas that are blood-red in colour. Silica-rich fluids circulated through the cooling lava. Agate precipitated from the fluids when they cooled sufficiently.

4. Monadh Dubh, north of Guirdil Bay: the Torridonian sandstone here is overlain by much younger sandstones of Triassic age. The contact between the two represents a mind-blowing gap in the geological record of around 650 million years.

Eigg

5. Sgurr of Eigg: much admired since early nineteenth-century adventurers visited the island. The complex geology is described on pages 31–2, but just stand back and admire the view. It's one of the most iconic in Scotland.

6. Eigg's east coast: walk in the footsteps of Hugh Miller, the famous geologist and writer who visited the island in 1845. The Jurassic rocks of Eigg have been on geologists' bucket lists ever since. North of Kildonan is perhaps the best access point. The path to the coast is steep in places.

Canna

7. The cliffs of Dun More, Dun Beag and Sanday: Sedimentary layers of conglomerate that lie between the lava flows are best seen here. These deposits were formed as eruption of lavas temporarily ceased and extensive rivers flowed across the land, dumping sandstones, pebbles and boulders eroded from the surrounding rocks.

Muck

8. Camas Mor: Jurassic rocks are exposed on the south coast. It's an interesting juxtaposition of lavas and Jurassic sediments. A thick gabbro intrusion has thermally metamorphosed (or cooked and thus altered) the Jurassic limestones into which it has come into contact. It is best seen along the east side of the bay.

East coast of Eigg, with Jurassic rocks in the foreground and lavas forming in the inland cliffs.

Acknowledgements and picture credits

Thanks are due to Professor Stuart Monro OBE FRSE and Moira McKirdy MBE for their comment and suggestions on the various drafts of this book. I also thank Debs Warner, Mairi Sutherland, Andrew Simmons and Hugh Andrew from Birlinn Ltd for their support and direction. Mark Blackadder's book design is up to his usual very high standard. Scottish Natural Heritage, in association with the British Geological Survey, published the *Landscape Fashioned by Geology* series that was the precursor to the new *Landscapes in Stone* titles. I thank them both for their permission to use some of the original artwork and photography in this book. Kathryn Goodenough and Tom Bradwell from BGS wrote the original text. This book completes the national coverage of thirteen titles in the *Landscapes in Stone* series. I must give special thanks to Hugh Andrew, Birlinn's MD, for his unswerving commitment to this task. I also acknowledge the James Hutton Foundation's financial support in publishing this title. I dedicate this book to Lorne Gill. He was a colleague at Scottish Natural Heritage, recently rebadged as NatureScot, for many years. His award-winning photographs have decorated every book I've written and all the illustrated talks I've given. Many of his scenic shots grace this book, too.

Picture credits

2–3 John Potter/Alamy Stock Photo; 6 Arthur Campbell/Shutterstock; 10 Helen Stirling (map); 12 drawn by Robert Nelmes; 13 drawn by Robert Nelmes; 14 drawn by Jim Lewis; 16 Lorne Gill/SNH; 17 drawn by Jim Lewis; 18 Lorne Gill/SNH; 20 illustration and photograph, reproduced with permission, Elsa Panciroli; 21 ansharphoto/Shutterstock; 22 drawn by Jim Lewis; 24 Helen Stirling (diagram); 27 Lorne Gill/SNH; 28 (upper) Lorne Gill/SNH; 29 Patricia and Angus Macdonald/Aerographica; 30 Lorne Gill/SNH; 31 Coatesy/Shutterstock; 32: Lorne Gill/SNH; 33 BGS; 34 Ondrej Prochazka/Shutterstock; 37 (upper) Rochard A McMillin/Shutterstock, (lower) Lorne Gill/SNH; 38 Lorne Gill/SNH; 40 Lorne Gill/SNH; 41 Lorne Gill/SNH; 42 Skye Studio LK/Shutterstock; 43 TravellingFatman/Shutterstock; 44 Helen Stirling (map); 45: Arthur Campbell/Shutterstock; 47 trevor hunter/Alamy Stock Photo.